Cutters and Crushers

Claire Llewellyn

KAEDEN BOOKS™

Contents

All sorts of teeth

Jaws and teeth are very important – where would animals be without them? Animals use their jaws first to grasp their food, and then they use their teeth to cut and crush it into bits. This makes food easier to swallow and easier for the body to digest. Some animals' jaws are broad and powerful; others are dainty and small. Animals' teeth vary, too – from the long and spiky to the short and blunt. Why are they all so different?

Tooth Talk

Birds need to be light in order to fly. They don't have heavy jaws with teeth. They have horny beaks.

All kinds of food

Different animals live in particular **habitats** and eat particular foods. Over millions of years, their jaws and teeth have changed to suit their diets. This slow change is called evolution. Animal groups evolved in different ways. **Mammals** evolved a variety of teeth in their jaws, while the teeth of fish and **reptiles** are all the same.

This book looks inside animals' mouths – from the nut-loving squirrel to the deadly shark – and describes how each one's jaws and teeth deal with its chosen diet.

Animals eat many different kinds of food.

What about you?

Do you have a pet, or know someone who does? What kind of food does it eat? Does it have teeth? Ask an adult to help you take a look.

Fruit, leaves and grass

Plants are an important food for many animals. Like all foods, they need to be broken down so that the body can digest them. Soft fruits need just a gentle crushing. Nuts and **grains** have a hard outer case, which can be opened by gnawing or cracking. Grasses, leaves and bark are hard to break down, and need a lot of chewing!

Word stretcher

An animal that feeds on plants is called a **herbivore**.

Case study

The horse

Horses feed mainly on grass. They cut the grass with their sharp, front teeth. These are called the incisors. Then they chew the grass on their blunt, back teeth. These are known as the molars. The molars are flat but covered in ridges. As the horse chews from side to side, the ridges grind the grass into a soggy pulp, which can then be swallowed.

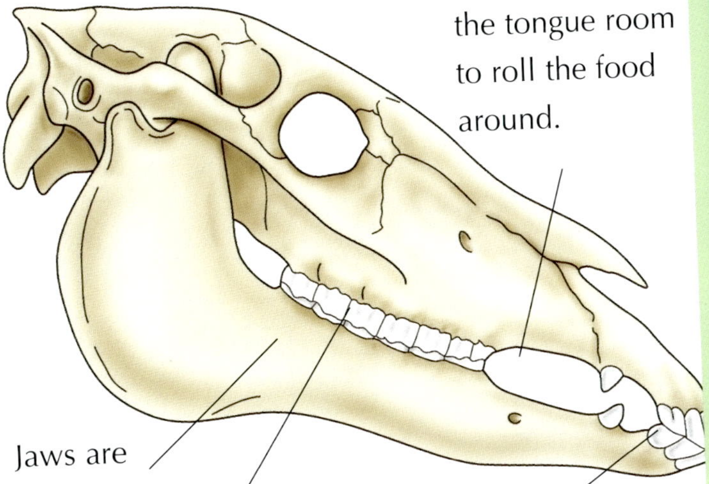

The skull of an animal shows its jaws and teeth. This is the skull of a horse.

What about you?

You eat lots of plant foods, too. Which teeth do you use to bite into an apple? Which teeth mash it to a pulp?

How to eat plants

The squirrel

Case study

A squirrel's skull

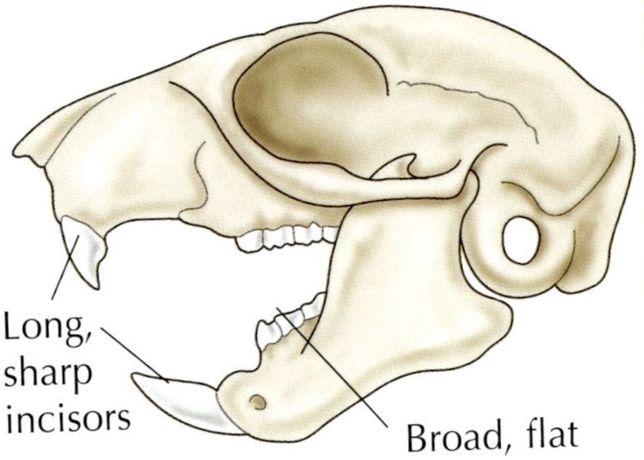

Squirrels are plant-eaters. They feed on nuts, fruit, buds and bark. Their incisors are sharp and very large – the perfect tools for cracking nuts and gnawing the **kernels** *inside. A squirrel's incisors never stop growing. If they did, they would soon wear out! Many animals have incisors like this. They all belong to a group called* **rodents**.

Tooth Talk

If a squirrel stopped eating, its incisors would grow so long that they would curl up around its skull.

A squirrel holds a nut in its two front paws while it gets to work with its incisors.

The elephant

Case study

An elephant's skull

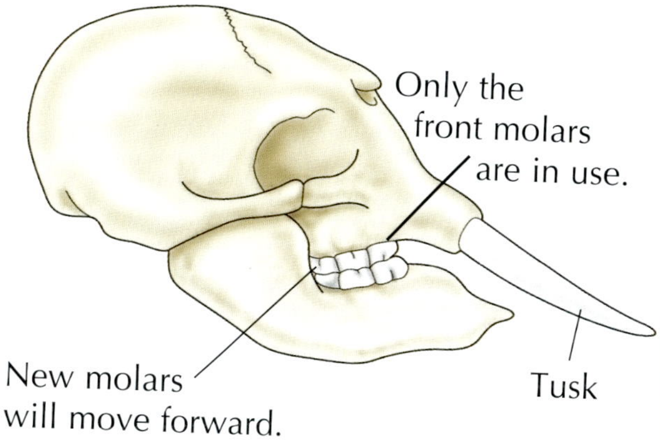

Elephants eat the leaves and branches of trees. This is tough food that needs a lot of chewing. The animal's jaws are lined with broad, flat molars, but only four teeth are used at a time – those at the front of the upper and lower jaws. In time these teeth are worn away, and new molars take their place.

Tooth Talk

An elephant's molar is the size of a brick and weighs 9 pounds – the same as four bags of sugar!

These animals also feed on plants.

Fancy some fish?

Many animals feed on fish. Fish are tasty and full of goodness, but they have slippery, **streamlined** bodies and can move very fast. How do animals catch hold of them? Some animals suck them into their jaws. Others pounce on them with snapping jaws and grasp them in sharp, spiky teeth.

Some sea foods, such as mussels and crabs, are protected by tough shells. Animals need strong jaws and teeth to crack the shell and reach the soft meat inside.

Word stretcher

An animal that eats only fish is called **a piscivore**.

What about you?

Do you eat slithery, slippery spaghetti? How do you grab hold of it? Which teeth, if any, do you use?

The dolphin

A dolphin's jaws are shaped like a long, narrow beak. They sweep easily through the water to snap at passing fish and squid. Inside the jaws are peg-like teeth, which grasp the live, slippery prey. There are no flat teeth to crush the food: the dolphin simply swallows it whole.

A dolphin's skull

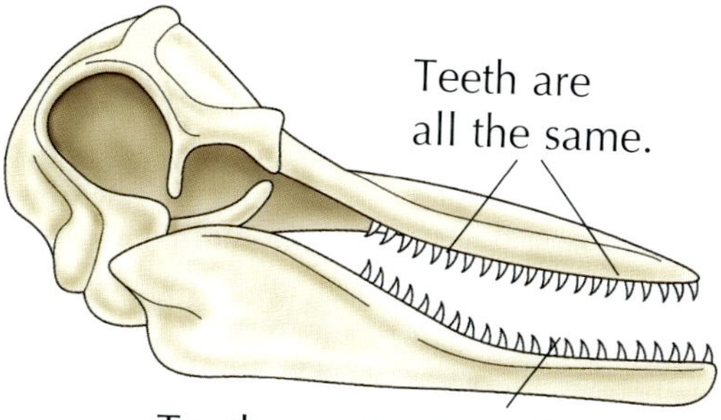

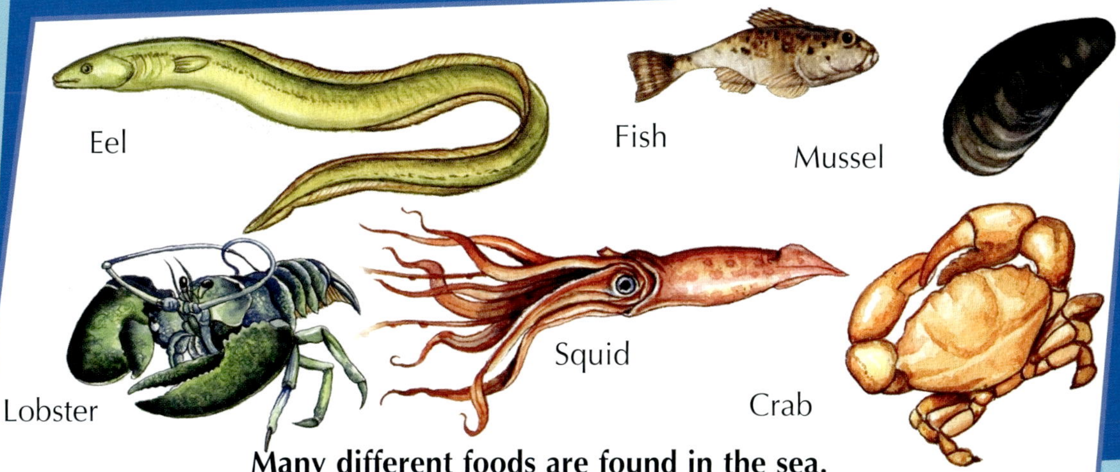

Many different foods are found in the sea.

Looking at fish-eaters

Case study

The eagle ray

Rays are large fish with fins like wings. Some rays are found deep down near the seabed, where they search for mussels and crabs. A ray's mouth lies underneath its body and contains strong, flat plates of teeth. These plates are good at crushing shells, which would break long, spiky teeth.

The skull of an eagle ray

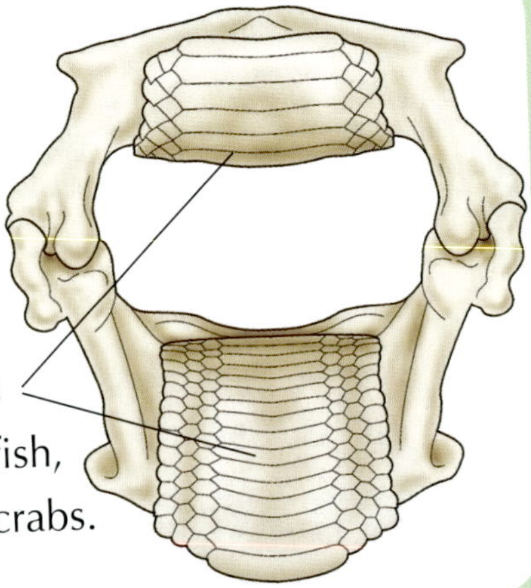

Case study

The piranha

Piranhas live in rivers in South America. These fish grow to no more than 23.6 inches long, but they have a fierce reputation. Their powerful jaws are lined with sharp, triangular teeth that fit perfectly together. With each snap, the prey is cut cleanly into chunks.

A piranha's skull

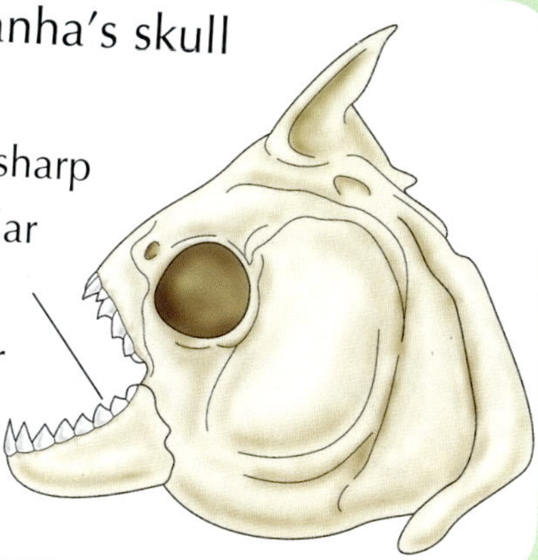

Tooth Talk

Piranha teeth are like razor blades: you could use one to sharpen your pencil!

These animals also feed on fish.

Meat on the menu

Many animals are hunters and feed on meat. Hunters have to catch every meal they eat, and need strong jaws and sharp teeth to make a kill. Then they have to eat the lifeless prey. Their teeth must bite, tear and chew the raw meat, and crack open the juicy bones.

What about you?

Do you sometimes eat meat off the bone? Which teeth do you use to pull the meat? Which ones chew it up?

Word stretcher

An animal that feeds on meat is called a **carnivore**.

Tooth Talk

Pet cats are meat-eaters. They have the same sort of teeth as tigers (opposite), and they use them in similar ways.

Case study

The tiger

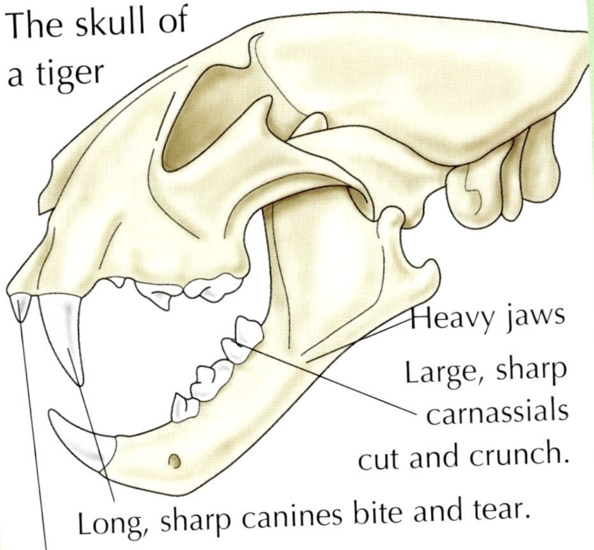

A tiger hunts antelope, pigs and deer. It bites its prey with its powerful jaws and stabs it with its dagger-like teeth. These lie at the corners of the mouth, and are called the canines. The canines work with the incisors to tear meat from the bone. Behind are larger teeth called the carnassials, whose sharp edges cut through hunks of meat and crunch through the toughest bones.

Beware: meat-eaters!

The snake

Case study

Snakes have remarkable jaws and teeth. Some snakes kill their prey with poison, which they inject through teeth called fangs. The fangs are hollow, sharp and long. They fold away inside the snake's mouth but spring forward when it attacks.

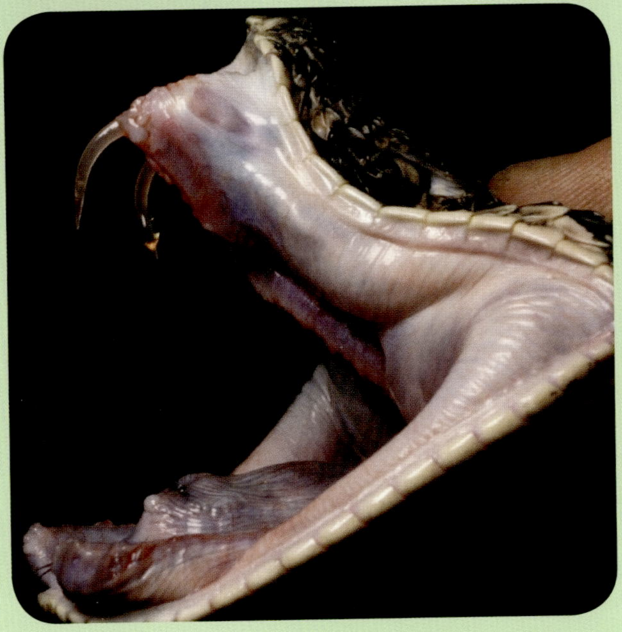

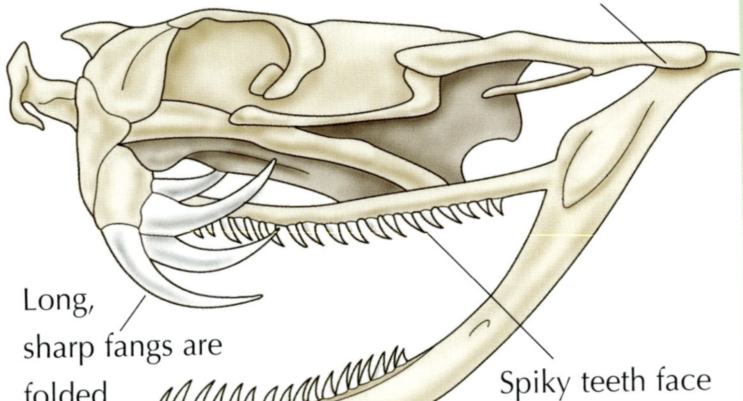

A snake's jaws help it to eat meals much bigger than itself. The upper and lower jaws unhook at the sides, giving the snake extra room.
It can stretch its mouth right over its prey and, inch by inch, swallow it whole.

Case study

The shark

The Great White shark has massive jaws and several rows of sharp, jagged teeth. With one bite, it can cut a seal in half. Sharks have no molars to chew their food. Instead, they grip their prey tightly and shake it wildly, while their teeth "saw" the meat into bits.

The skull of a Great White shark

Rows of new teeth replace teeth that are lost.

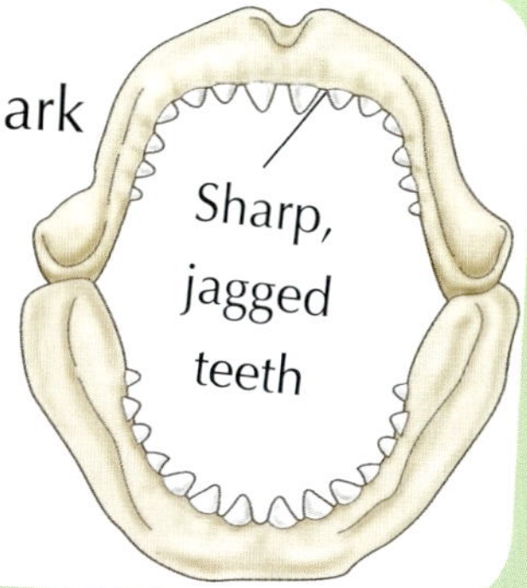

Tooth Talk

A shark loses teeth in every attack, but new ones move forward to fill the gaps. Some sharks get through thousands of teeth in their lifetime.

These animals are both meat-eaters.

A mixed bag

Some animals have evolved to eat a mixed diet. They will eat anything from small, juicy berries to gristly chunks of meat. These animals have developed a variety of teeth that can tackle all sorts of jobs.

The bear

Case study

Bears eat whatever food they can find – berries, honey, fresh grass, snails, fish, rodents or young deer. A bear's jaws contain a toolkit of teeth: sharp incisors for cutting and snipping, pointed canines for stabbing and tearing, and broad flat molars for crushing and grinding.

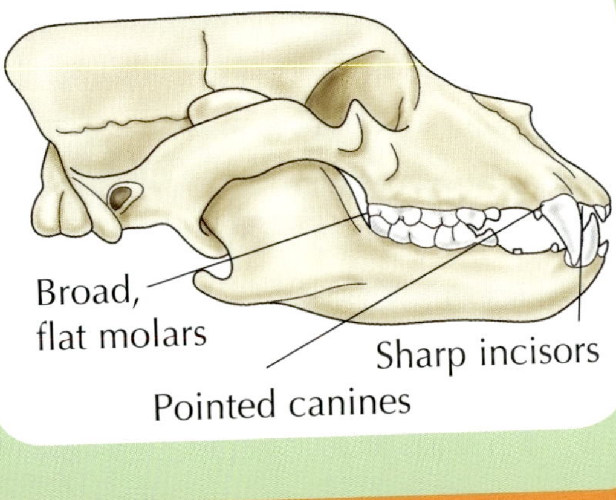

These animals also eat a mixed diet.

Word stretcher

An animal that eats both plants and meat is called an **omnivore**.

What about you?

Are you a herbivore, a carnivore or an omnivore? How many kinds of teeth do you have?

Tooth Talk

Vegetarians have the same teeth as people who eat meat. Teeth evolve very slowly, over many millions of years.

How teeth grow

Most animals' teeth are very hard-wearing. With luck, they will last a lifetime!

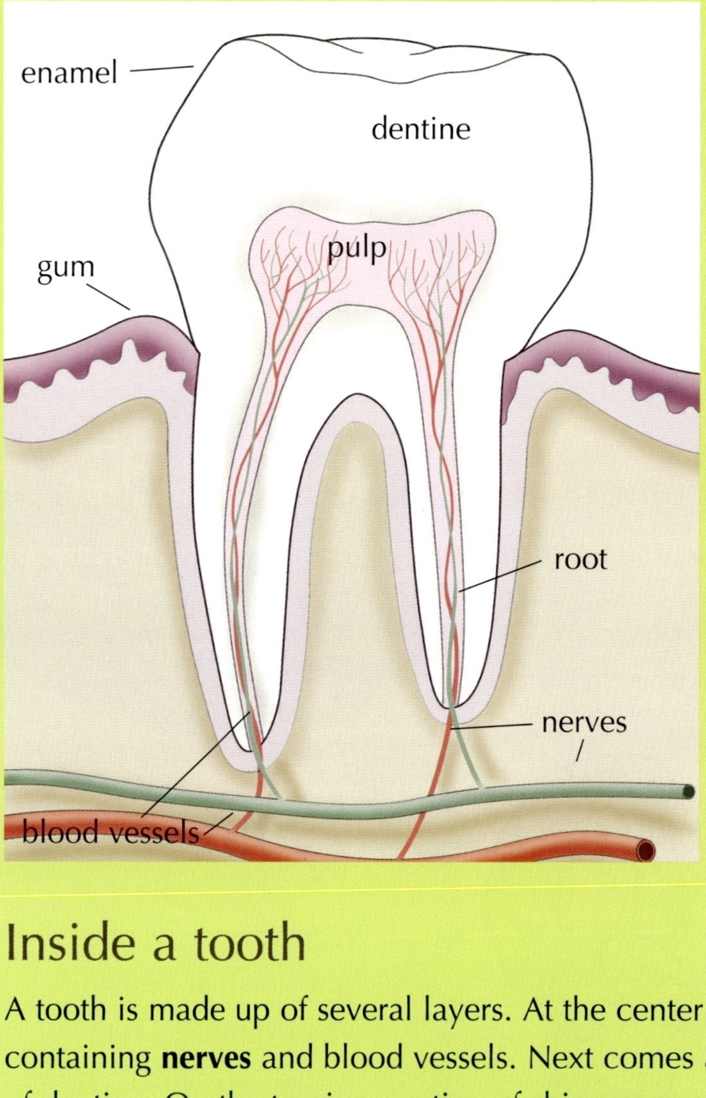

Tooth Talk

Teeth are slow to rot away. Teeth that once belonged to dinosaurs are sometimes found as fossils in the ground.

Inside a tooth

A tooth is made up of several layers. At the center is a soft pulp containing **nerves** and blood vessels. Next comes a hard layer made of dentine. On the top is a coating of shiny enamel, the hardest material in the body. The roots of each tooth grow deep into the jaw, holding it firm like the roots of a tree.

New teeth

As an animal grows, its jaws get bigger but its teeth do not. One by one, the teeth drop out and are replaced by larger ones. If an animal loses these teeth, they are lost forever. But some animals, such as crocodiles and sharks, develop lots of extra teeth to replace the ones they lose.

What about you?

Humans have a first set of 20 baby teeth, followed by a second set of 32 adult teeth. How many of each do you have?

Extraordinary teeth

Some animals have extraordinary teeth.
Did you know an elephant's **tusks** are really its teeth or that a whale's tooth grows in a spiral?

An African elephant's tusks are its two upper incisors that grow longer throughout its life. The elephant uses its tusks to frighten its enemies and dig roots from the ground.

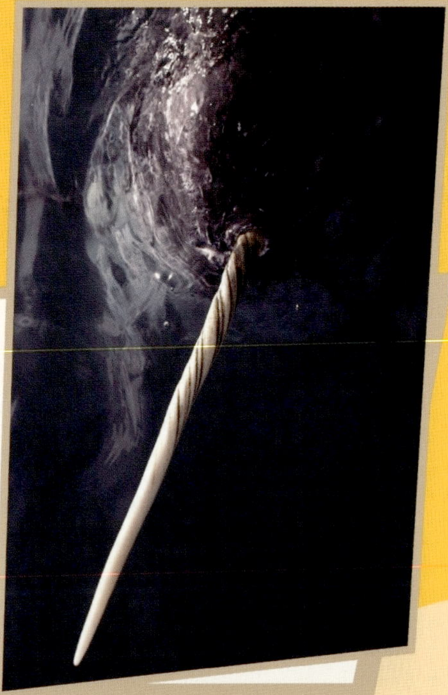

A male narwhal has a beautiful, spiraling tusk that grows out through a hole in its lip. What could it be for? Does it attract females, or scare off other males? Scientists still aren't sure.

A hippo's lower canines can be as long as your arm! These fierce animals know how to use their teeth, and have even bitten boats in two.

A walrus's tusks are useful tools. They help it to dig up shellfish from the seabed and to pull itself onto the ice. The male that has the longest tusks is usually the leader of the herd.

Glossary

fossil – the remains of an ancient plant or animal that have been turned into stone

grain – the seed of a cereal plant, such as wheat or rice

habitat – the natural home of an animal or plant

kernel – the inside of a seed or nut

mammal – a warm-blooded animal with a bony skeleton that feeds its young on milk. (Dogs, cats and mice are all mammals.)

nerve – the sensitive part of a tooth

reptile – a cold-blooded animal with a bony skeleton and a scaly skin. (Snakes, tortoises and crocodiles are all reptiles.)

rodent – a mammal with long, strong teeth that are adapted for gnawing. (Squirrels, rats and rabbits are rodents.)

streamlined – having a smooth body shape that moves easily through air or water

tusk – a long, curved, pointed tooth that grows out beyond the mouth. (An elephant has tusks.)

Index